The 250 Reminders for Managing Software Development Projects

By C.W. OH et al

I0464567

This page is intentionally left blank.

Table of Content

This page is intentionally left blank.

1. Introduction

Software development projects are complex endeavors that require careful planning, monitoring, execution and coordination with the different stakeholders. And in the midst of all these activities, we sometimes forget how best we can handle software development projects. And this is where the book comes in.

While we get ourselves involved in the details of the day to day software development project activities, this book intends to bring us a step back to reflect and to rethink on the strategy of the software development project at hand, and what we can do more to improve on the current status of the project.

The book does so by presenting individual reminders, that cover all areas on software development projects; project startup, requirements, design, implementation, delivery, acceptance. These are presented such that readers can easily understand, what needs to be done, what are the common and good practices, what are the potential pitfalls, etc.

These reminders are generally divided into phases that a standard software development project will undergo, and similar to a standard waterfall model approach. And the reminders are generally oriented towards the deliveries of the project and any other activities that will contribute to the deliveries.

It is the hope of the authors that by reading these reminders, they give readers a different perspective on software development projects, so that when the readers go back to work on the project again, they are able to do reviews in their specific areas better and to rectify any potential faults and implement corrective actions where needed.

This book is intended to be an easy to read reference for IT professionals or any persons involved in software development projects. The book can be read in different ways, i.e. from the start to the end, or directly to the specific reminder.

To maximize the content of the book, it is recommended that the book is first read from the start to the end, to familiarize the contents available, after which the book can be access directly to the reminder when needed.

***

2. The 250 Reminders

2.1 Project Startup

Reminder #1:

Software development projects are non-trivial undertaking. There must be serious thoughts, planning, re-planning, reviews, revisions, monitoring and actions to ensure that software development projects do not go terribly wrong. Constant efforts are needed to push all things forward until completion.

***

Reminder #2:

Good and well-defined processes are important and they are the keys to successful software development projects.

***

Reminder #3:

Do not hesitate to invest the project time, effort, and budget to inculcate good processes, especially at the start of the project.

***

Reminder #4:

If there are processes that do not work, review them immediately. Understand why these processes are not working, revise them, and implement the revised processes as soon as possible.

***

Reminder #5:

If there is no process defined or documented, do not let this pass. You can start by writing down the current, processes, and improve from there.

***

Reminder #6:

You need to have a software development plan. This is the plan that says how you will develop the software and how you will deliver the software. The plan usually includes the software development project description, business needs, manpower plan, communication plan, quality plan, change control & configuration plan, validation and verification plan, documentation plan, delivery plan, and risk plan.

***

Reminder #7:

The different sections of the software development plan, e.g. configuration plan, quality plan, test plan, these can be written separately. But the software development plan needs to refer these separate plans.

***

Reminder #8:

The software development project starts with a project charter. It is the mission statement of the project, and the document authorizes the existence of the project.

***

Reminder #9:

People are your best asset in any software development projects. Each of them contributes their expertise to the project. They are professionals of their domain. Their contributions to the software development project are essential. Without the people, there will be no software development project.

***

Reminder #10:

The project team needs to be convinced that the processes are there to help and that the processes that they follow move them closer to achieving the project objectives. They need to be convinced because they are the ones that will be following the processes. If they are not convinced, they will probably not practice them. And this defeat the purpose of having processes in the first place.

***

Reminder #11:

It is essential that all the project team members are convinced that the project is achievable. They need to be educated about the objectives of the project and how these can be achieved.

***

Reminder #12:

The project team members can sometimes be located across different locations. Make use of video conference and instant messaging to keep in contact. And have regular remote meetings.

***

Reminder #13:

Where possible, locate your project team together in the same office, where effective work can be best achieved. The face-to-face interaction also helps in building team relationships and improves communication between the project team members.

***

Reminder #14:

You must have a project organization chart. It tells who the people in the project team are, who is responsible for what and who is reporting to whom. It keeps everyone in their allocated place within the project.

***

Reminder #15:

Start your software development project with a group of small senior staff, e.g. software architect, test lead, system engineering lead, quality assurance & configuration lead. Expand the team the project progresses and when the project requires the additional manpower.

***

Reminder #16:

With the senior team members, invest your time and effort in getting the processes right and the communications between the different team at the start of the project.

***

Reminder #17:

Choose your next layer below, or your team leads, carefully. Pick the best person for the job. Remember, you will need to work closely with them throughout the duration of the software development project. They must be able to work with you and they must be supportive of your plans and efforts.

***

Reminder #18:

Do not micro-manage your team members. You can help them by asking them how they are doing, giving them your feedback, and ask if they need help. If necessary, augment them with experts to help them along.

***

Reminder #19:

Identify the person in the project team that will be your second-in-charge, i.e. the person that make the decision, during your absence. This should be reflected in your project organization chart also. Announce this delegation prior to any absence.

***

Reminder #20:

Get you next layer below, i.e. your team leads to identify their second-in-charge also. Their second-in-charge should be reflected in your project organization chart also.

***

Reminder #21:

The project team members need to produce software for the project, and healthy team members at most efficient in delivering the software for the project; the needs of the team members must be addressed first before they can be productive to the project.

***

Reminder #22:

Understand Maslow's hierarchy of needs theory, and the different needs of any person; from Physiological, Safety, Belonging, Esteem, to Self-Actualization.

***

Reminder #23:

When building your project team, you want to foster team spirit and team work. You need to understand Tuckman's different stages of team formation; where it goes from:

(1) Forming,

(2) Storming,

(3) Norming and

(4) Performing.

***

Reminder #24:

Do not play favorites amongst your team members. You may unknowingly fragment the team by your actions. Within the project team, there will always be some that are better than others. What is more important is that you need them to work together as a project team to achieve the project objectives.

***

Reminder #25:

Show your sincere appreciation of the work of your project team members. Learn the different methods of boosting the staff morale. A happy team goes a long way.

***

Reminder #26:

Give your software developers the best place to work. They need the quiet and conducive place to do their work properly because the work that they perform is generally complex and requires a high level and prolonged concentration.

***

Reminder #27:

Assign software developers in work that are close to their interest. Software developers will find their work more interesting, and they will be highly motivated and productive.

***

Reminder #28:

When dealing with people, remember Emotional Intelligence (EQ), i.e. the ability to recognize one's emotion as well as recognizing the other person's emotion in order to adapt and guide one's behavior and thinking in a positive direction.

***

Reminder #29:

Project stakeholders are people and organizations that are affected by the software development project. They may be internal or external to the project, and their interests can impact positively or negatively on the project. Understanding who your stakeholders are and who will impact your project positively and negatively is important.

***

Reminder #30:

Select and use project management software to manage your projects, e.g. Primavera, Clarity, MS-Project, ProjectLibre, GanttProject, etc. Make sure that the software that you picked suits your project and organization information needs and is easy to use.

***

Reminder #31:

When using the project management software to create a project, take note that you have these completed:

(1) Set the calendar and working hours,

(2) Identify the public holidays,

(3) Identify the leave days that the project team members are taking,

(4) Enter the tasks and durations,

(5) There should be a start milestone and an end milestone,

(5) Sequence the tasks using dependencies between the tasks,

(6) Enter your resources,

(8) Assign your resources to the tasks,

(9) Check that the resources are not over allocated.

***

Reminder #32:

WBS stands for Work Breakdown Structure.

***

Reminder #33:

When listing your individual tasks of your project, these must be broken down into no more than three (3) days of work, or no more than one percent (1%) of the total project effort.

***

Reminder #34:

You must identify and know your critical path of your project plan at any time during the project, and make sure that the activities that are on the critical path do not get delayed.

***

Reminder #35:

The critical path of a project is the sequence of tasks in the project such that any delays on any of these tasks, the project will be delayed accordingly, i.e. if one of the tasks is delayed by one day, then the project will be delayed by one day also, or zero slack.

***

Reminder #36:

In project management, slack is the amount of time that a task that can be delayed without delaying the overall project schedule.

***

Reminder #37:

Share the project plan with everyone in the project so that everyone knows how the software development project will be executed and how everything fits together. Let the project team knows who needs to do what tasks and when they must complete those tasks.

***

Reminder #38:

All plans are subjected to changes and revision. As the project progresses, keep your project plan updated on a weekly basis and revise your project plan when necessary.

***

Reminder #39:

Once your project schedule is approved, remember to baseline your project plan so that you have a reference to check what has deviated from the original plan. Most project management software has this baseline function.

***

Reminder #40:

Make use of the progress line function in most project management software to easily identify those tasks that are overdue as of current date of checking.

***

Reminder #41:

Have check points inside your project plan, and conduct the checks to ensure that all things are performing as planned.

***

Reminder #42:

You can utilize an issue/defect tracking software to automate the tracking of your software development project development activities, where the tasks in the project plan are registered as issues in the issue/defect tracking software, which are than assigned to the respective project members to perform and complete. The issues are then updated by the team members on the issue/defect tracking software. From here, we will be able to check the status of the project from the issue/defect tracking software.

***

Reminder #43:

Have a project website or a project status board so that everyone is kept updated on the progresses.

***

Reminder #44:

The project website or the project status board should minimally show:

(1) The schedule of the project,

(2) The milestones of the project,

(3) Show recently completed tasks,

(4) The next steps,

(5) Issues,

(6) Defect statistics, and

(7) Top ten risks.

***

Reminder #45:

Identify the different metrics and Key Performance Indicator (KPI), e.g. earned value scheduled indicators, cost and budget, tasks status, etc that you want to measure for the software development project. These metrics should be relevant and are good indicators of what you want to track of your software development project.

***

Reminder #46:

"What gets measured gets managed" – Peter Drucker.

***

Reminder #47:

The project manager must control the progress of his/her software development project. Controlling the project is better than the project being out-of-control.

***

Reminder #48:

The project manager is like the conductor of a musical orchestra. He orchestrates all the different players in the project team to work together to complete the project objectives. As such he organizes plans, schedules, monitors, reports, etc. He/she gets things moving and makes things happen

***

Reminder #49:

The project manager is in charge and he/she is responsible for the project success and failure.

***

Reminder #50:

Use Earned Value Method is a technique that can be used to measure the progress of a software development project in terms of cost and schedule. It can give an objective view on the status of a software development project.

***

Reminder #51:

A project repository is a central place all project related items are kept. Physical objects can be kept in a file cabinet, and soft copies, they can be stored in a shared folder, or on the project website.

***

Reminder #52:

Scan a copy of the physical items, e.g. contract, project documents, etc, and upload them onto the shared folder. Consider how to access to the project repository at remote locations. The scanned copies can then be access easily when needed.

***

Reminder #53:

Have a standardized storage methodology for the project repository so that you and your team can easily find the documents that you need.

***

Reminder #54:

Organize and hold regular project team meetings on project status updates, issues, and action plans. And to make these meetings productive, standardize the project meeting agenda and learn how to organize effect meetings.

***

Reminder #55:

Keep your customers updated. Organize and hold regular meetings with your customers on project status, issues, and action plans.

***

Reminder #56:

Don't be afraid to communicate bad news. When communicating bad news, take a proactive approach, mention the cause of failure, and more importantly suggest action plans to recover and improve the situation.

***

Reminder #57:

MOM stands for minutes of meeting. The minutes of meeting serves as a record of the any decision made at a meeting and the action items agreed. Assign someone to take down the minutes of meeting at every meeting.

***

Reminder #58:

Action items in the minutes of meeting must have the description of what to do, who is responsible and performing the action item, and by when. Do not assign any action item to a group of people, otherwise no one will feel responsible for it.

***

Reminder #59:

Make sure that you follow up on the action items in the minutes of meeting. Chase after the people who are responsible for the action item for status update. Any action item that is not closed needs attention.

***

Reminder #60:

Standardize your project reporting, and have a reporting template that can be used for every project reporting. Make use of the metrics and key performance indicators as part of your reporting to show the current status of the project.

***

Reminder #61:

The reporting template should cover the following areas:

(1) Overall schedule and milestones,

(2) Activities,

(3) Issues,

(4) Risks,

(5) Opportunities,

(6) Financials,

(7) Quality assurance, and

(8) Manpower.

***

Reminder #62:

Most of the cost in a software development projects comes from manpower. And hence be careful in managing your manpower resources

***

Reminder #63:

Categorize the time charge codes properly at the start of the project. The categories generally reflect your work breakdown structure, e.g. Project Management, Quality Assurance, Requirement, Design, Implementation, Delivery, Acceptance, and Maintenance. Otherwise, create time charge codes for those activities that want keep track.

***

Reminder #64:

Use a timesheet tracking software to track your project time charges. This makes the consolidation of your time charges simpler. The timesheet tracking software should have a function to allow you to check and approve any time charge submitted.

***

Reminder #65:

To track rework, the project will need the relevant rework time charge codes.

***

Reminder #66:

When the time charge codes are ready and the timesheet tracking software is ready to use, organize a meeting with the project team members and educate the project team how and where to time charge their activities to.

***

Reminder #67:

All software development projects have budgets. Divide your software development project budget down into different areas. This should minimally be the project budget, project contingencies budget, and management contingencies budget. The project contingencies budget is the monies for the known risks, and management contingencies budget is for any other risks that is unknown.

***

Reminder #68:

Changes are inevitable in the project, but they must be managed. Setup a change control board to manage all the changes in the software development project. All change proposals are to be submitted to the change control board for decision. The change control board should consist of people from the project team and from the customer.

***

Reminder #69:

Standardize the change control proposal document template for the project, and all change proposals are to be submitted using the standardized template.

***

2.2 Requirement

Reminder #70:

Some characteristics of good requirements:

(1) Complete,

(2) Clear,

(3) Current,

(4) Consistent,

(5) Feasible,

(6) Unambiguous,

(7) Prioritized,

(8) Verifiable,

(9) Traceable,

(10) Necessary,

(11) Implementation free.

***

Reminder #71:

Look up on RFC 2119 – Keywords for use in RFCs to indicate Requirement Levels, where it provides guidelines for keywords when writing requirements, e.g. MUST, MUST NOT, SHALL, SHALL NOT, SHOULD, SHOULD NOT, RECOMMENDED, MAY and OPTIONAL.

***

Reminder #72:

Software functionalities are inherently intangible. You need to have one or many test cases to proof that your software has achieved the software functionality defined. And you need a traceability matrix to trace all the software functionalities defined to all the test cases.

***

Reminder #73:

It is essential that you know what your software development project is expected to deliver, e.g. software, documents, reports, attendance of meetings, reviews, performance standards, etc. Have them listed, and check them off every time a delivery is delivered. You also need to know when you need to deliver them.

***

Reminder #74:

Use a requirement management software tool to keep track of the requirements of the software development project. Such software generally has traceability function that allows, for example to create links between requirements and test cases.

***

Reminder #75:

From your contract documents, identify all the keywords, e.g. SHALL, MUST, SHOULD, etc. The software development project needs to show that it has achieved these requirements by tracing the artifacts produced by the project back to these requirements

***

Reminder #76:

For tracing strategy, trace from

   (1) Contract to requirement,

   (2) Requirement to design,

   (3) Design to source-code,

   (4) Source-code to unit-test, and

   (5) Requirement to test-case.

If budget and time are constrained, (1) and (5) are the minimum.

***

Reminder #77:

There are different type types of requirements and some of which are:

(1) Functional,

(2) Non-functional,

(3) User interface,

(4) Reliability,

(5) Performance, and

(6) Capacity.

***

Reminder #78:

Get the end-users involved as early as possible in the project. They are the people who will eventually use the software that your team is building.

***

Reminder #79:

Sometimes the users do not know what they want; as a result they want everything. The project team needs to help and guide them to know what they need.

***

Reminder #80:

Make use of user interface prototype to further solicit and refine the requirements from the end users. This is especially useful when the requirements are unclear or incomplete.

***

Reminder #81:

Schedule user interface prototyping activities into your project schedule.

***

Reminder #82:

The user interface prototype should cover all, if not most, the screens of the software. When possible, it should mimic the behavior of the software.

***

Reminder #83:

Leverage on user interface prototype to reduce risks of the project. And use the user interface prototype together with a story board to describe the use of the software.

***

Reminder #84:

A storyboard, in this context, is a sequence of pictures or drawings that describes the visual displays and the possible interaction of the user interface prototype.

***

Reminder #85:

Have a user interface style guide to standardize the look and feel of the user interface. The theme should be consistent and reflect the purpose of the software.

***

Reminder #86:

In concurrent with the user interface prototyping, the user manual can be developed and checked with the end-users on the functionalities and expectations of the software.

***

Reminder #87:

Having the user interface prototype and the user manual manage the end-user expectations, bring familiarity and acceptance of eventual software.

***

Reminder #88:

Draw a context diagram of your software system, to show and understand the inputs to the software and the outputs of the software.

***

Reminder #89:

Expand on the context diagram, to show the data flows, as inputs to the software, within the software, including data storage, and outputs of the software. Where necessary, break down the data flows down into several levels, i.e. data flow diagram level 1, level 2, and level 3 and so on. The emphasis should be on the data flows.

***

Reminder #90:

Entity relationship diagram (ERD) is a way to describe a process, where the entities (or objects) are associated and dependent with each other by means of relations, e.g. a staff must belong to a department in the company.

***

Reminder #91:

Use case diagram is a way to describe the interactions between a user and the software, with each interaction as a use case.

***

Reminder #92:

At the end of the requirement phase, the project should have at least a system/sub system document that contains the requirements to be met by the software. These requirements must be able to back to the contract.

***

Reminder #93:

Do not proceed on to the next phase if the requirements are not completely captured, or any other deliverables associated with the requirement phase are not completed.

***

Reminder #94:

Fixing requirement defects or ambiguities early in the project reduces downstream cost, e.g. one requirement can results several design documents, several test cases, a few hundred lines of codes, and many pages of user documents. So if one requirement changes at the later stage, all these other artifacts need changes also.

***

Reminder #95:

Conduct requirements reviews, internally and with your customers.

***

Reminder #96:

Highlight any deviation of scope the contract as soon as possible, and record down these variations. Make us of the project change control process and the change control board to manage these variations.

***

Reminder #97:

Changes to requirements happen. Make use of the project change control process and the change control board to manage any requirement changes.

***

Reminder #98:

Ensure that all requirement defects and revisions are corrected and completed before moving on to the next phase.

***

Reminder #99:

Before moving on to the next phase, conduct reviews the project, i.e. plans, cost, schedule, risks, and perform re-estimation.

***

2.3 Design

Reminder #100:

The design phase is the time to describe how the software will be implemented, including what technologies to use and how these will be used, e.g. development and design tools, databases, middleware, and any other third party applications needed.

***

Reminder #101:

In general, the design documents, when completed, must be in such details, such that they are in a suitable form for the programmers to perform coding. They must contain further elaboration of the software down to each of its components, where the algorithms in pseudo-codes of routines/classes are described also, i.e. from design to codes

***

Reminder#102:

All data fields used by the software documented, explained and described in the data dictionary.

***

Reminder #103:

All data flows and their data types must be able to be described, from entry to the software, used in routines, stored in the database, retrieved from database, output to external of the software. In addition, the technologies to transfer, store and retrieve the data need elaboration also.

***

Reminder #104:

The size of the data can be estimated here given the data fields used and stored. This estimation can give the possible amount of data storage needed, and the performance of the hardware needed.

***

Reminder #105:

The behavior of the software must be estimated based on the possible usage scenarios, in terms of performance, reliability, data size.

***

Reminder #106:

The project will need to trace from the requirements to the design to proof that the design covers all requirements.

***

Reminder #107:

Read and understand the good practices of good software design. In addition, discover and trial the available software design tools that might be suitable to your needs.

***

Reminder #108:

Continue with the user interface prototype to get the approval and confidence of the customers. Further elaborate the user interface prototype sufficiently to show the look and feel of the software, and to be able to use it to write the user manual.

***

Reminder #109:

User manuals can help the customer understand how the final software will work. These can also be used as supporting documents for the project team to understand what software they will be building.

***

Reminder #120:

This is the phase where the project team needs to define the hardware specifications, e.g. process speed, number of processes, random access memory (RAM), hard-disks space, screen size, etc for the software.

***

Reminder #121:

Does the software require the use of a database? If yes, what is the preferred database application to be used together with the software?

***

Reminder #122:

Does the software need a middleware? What middleware application is preferred?

***

Reminder #123:

Is the software standalone, client-server architecture, or web-based? If your software a web-based, what web-server application is preferred?

***

Reminder #124:

Platforms and technology considerations: Windows, Linux, or others? .NET, Java, or others?

***

Reminder #125:

Maximize your hardware, and make use of virtualization technologies to emulate the hardware.

***

Reminder #126:

This is the time where the project team needs to decide on selecting the appropriate software development tools. Standardize these tools across all the programmers.

***

Reminder #127:

Think technology around business processes, and not the other way around.

***

Reminder #128:

Conduct software design reviews, internally and with your customers. If the design reviews are lengthy, spread out the reviews into several sessions.

***

Reminder #129:

Invite technical experts, external of the project team, to perform audits on the design. These audits can give external insights on the adequacy and robustness of the software design.

***

Reminder #130:

Do not proceed on to the next phase if the designs are not, or any other deliverables associated with the design phase are not completed.

***

Reminder #131:

Ensure that all design defects and revisions are corrected and completed before moving on to the next phase.

***

Reminder #132:

Before moving on to the next phase, conduct reviews on the project, i.e. plans, cost, schedule, risks, and perform re-estimation.

***

2.4 Implementation

Reminder #133:

The project team needs to maintain the different software environments for different purposes. They are:

(1) Development,

(2) Integration,

(3) Staging,

(4) Acceptance & Testing and

(5) Operational.

***

Reminder #134:

If the five environments are a bit too much for your software development project team to maintain, the minimum must have environments are:

(1) Development,

(2) Integration, and

(3) Operational.

***

Reminder #135:

The progressions of the software developed, across the different environments, are as follows, starting from left to right, and top to bottom:

Development → Integration → Staging → Acceptance Testing → Operational

***

Reminder #136:

The development, integration and staging environments are generally located at the office where the software development project team is located. The acceptance testing and operational environments can be located at the customer site. If there is no acceptance testing environment available, the staging environment can be used as one instead.

***

Reminder #137:

When a software code moves from one environment to another, e.g. from development to integration or integration to staging, it must undergo and pass the tests and reviews. This process must be strongly enforced.

***

Reminder #138:

The development environment is where the developers do their programming, debugging of their codes and unit testing. This environment is generally maintained by the developers themselves.

***

Reminder #139:

The integration environment is where all the developers' produced applications are integrated and tested. This integration environment must be maintained by a separate software integration engineer.

***

Reminder #140:

The staging environment contains the version of the software that has passed all the integration tests, and destined for the acceptance and operational environments.

***

Reminder #141:

At the staging environment, this is the place where the software will be kept constantly running, just like as in an operational environment, with the tasks and hope of identifying any software bugs that might have not been captured previously. This is a proactive measure of identifying and capturing software bugs early.

***

Reminder #142:

The staging environment, in terms of software version and system configurations, is the nearest to the operational environment that is located in the software development project office.

***

Reminder #143:

The development, integration and staging environments should be equipped with good and representative hardware and simulators to emulate the operational environment. It should also contain a variety of good debugging tools to capture software bugs.

***

Reminder #144:

The acceptance test environment, if created in the software development project, is an environment where the software application built by the software team will be tested by the customer. The software introduced into this environment should have already passed the integration tests, and completed robustness and duration tests at the previous environments.

***

Reminder #145:

The staging environment can be used as the acceptance test environment, if the acceptance test environment does not exists.

***

Reminder #146:

This acceptance test environment is generally located at the customer site and has, if not almost, the same system configurations as the operational environment.

***

Reminder #147:

The responsibility of maintaining the integration, staging and acceptance testing environments must be assigned to a responsible system engineer reporting to the test manager.

***

Reminder #148:

All tests performed at the acceptance test environment by the customers are generally considered as a formality. This is these tests should have already been conducted, and passed, prior to the acceptance test events with the customers.

***

Reminder #149:

The software development project team must define the process of managing and handling any bugs or errors discovered after development environment. And this should include the recovery turnaround time expected.

***

Reminder #150:

Any bugs or errors discovered after development environment must be fixed immediately, if not as soon as possible, by the responsible programmer.

***

Reminder #151:

Enable more and efficient testing; make use of automated software testing tools to automate testing. You can get a better and faster feedback on the health of your software.

***

Reminder #152:

Before the coding starts, develop or adopt coding standards for the project team. Educate all the programmers and the other project team members on the coding standards, its purpose and the need to adhere to these standards

***

Reminder #153:

Use automated source code version control software to maintain the versions of the source codes. Check that all the project team members know how to use it.

***

Reminder #154:

Focus on getting good, robust and reliable software developed and have it shipped to the customer.

***

Reminder #155:

Educate the project team to develop what is needed to complete the task, and do not gold plate your software functionalities.

***

Reminder #156:

Gold plating is defined as the continuation of tasks that is not within the expected requirements, e.g. enhancing the software with the thought of impressing customer with the additional features not listed in the original scope.

***

Reminder #157:

Instead of delivering all the software functionalities at one big delivery, have the software delivered in stages and develop the important functions first. In a staged delivery of software functionalities, the different functionalities are developed and delivered in stages, e.g. if the whole software has ten functions, the first three functions will be developed first, pushed to integration environment, followed by the next three, and then the next three and so on.

***

Reminder #158:

Develop a software integration plan. Have the integration plan reviewed by the project team, and by people external of the project team. Educate all the programmers and the rest of the project team members on the integration plan and how it impacts them.

***

Reminder #159:

Develop the individual processes for all the project team members, e.g. programmer processes, software tester processes, etc. Revise these processes when needed and keep these processes up to date.

***

Reminder #160:

The work assignments to the programmers, and to all other team members, must be detailed; the tasks, the standards, and completed by when. And these should be reviewed on a weekly basis.

***

Reminder #161:

Develop and implement a process/system to facilitate the creation tasks, assignment tasks, tasks update and task feedback. Constantly review and update this to keep this optimized.

***

Reminder #162:

Utilize software to automate the process/system to facilitate the creation tasks, assignment tasks, tasks update and task feedback, such that the information can be accessed easily from any workstations, or remotely.

***

Reminder #163:

Have you considered testing your software for code coverage?

***

Reminder #164:

Standardize all the software development tools, e.g. integrated development environment (IDE), compiler, operating system, debuggers, etc, and any other software tools used by the project team. Everyone in the team should be able to know how to use these tools.

***

Reminder #165:

Standardize all templates, e.g. document templates, unit test record template, source code templates, etc, and enforce the use of these templates. If there is no template available, start by creating one.

***

Reminder #166:

Implement a daily build and test practice where the software is build daily with the latest version, and tested. If there are anything broken due to the new introduction must be rectified immediately.

***

Reminder #167:

The software development project team must treat any commits of source code seriously and that they committed codes that are properly written, and that the code had be reviewed and tested extensively.

***

Reminder #168:

The software development project team must be reminded that quality work must be done at all times, instead of delaying at a later stage.

***

Reminder #169:

Do not allow the test later and fix mindset to propagate within the software development project team. Discourage your programmers from delaying or pushing the testing of their source code until software integration at the integration environment. They must be educated that they must product quality work at their levels.

***

Reminder #170:

Have some incentive system in place, for programmers, for zero bug or lowest number of bugs discovered during testing at integration, staging, acceptance testing and operational environment.

***

Reminder #171:

Likewise, have some incentive system in place, for software testers to search for bugs and errors, e.g. software tester having the highest number of bugs found.

***

Reminder #172:

All bugs and errors discovered after development environment, i.e. at integration environment and onwards, are to be recorded into the project issue/defect tracking software. These bugs and errors must be rectified immediately, if not as soon as possible.

***

Reminder #173:

If there are programmers having too many bugs, constantly, found at the integration environment and onwards, reviews must be conducted and have recovery plans in place. If necessary, assign a senior programmer to guide the programmer until the problem is resolved.

***

Reminder #174:

Do not proceed on to the next phase if the any other deliverables associated with the implementation phase are not completed. Ensure that all design defects and revisions are corrected and completed before moving on to the next phase.

***

Reminder #175:

By the end of the implementation phase, you should have at least working version of the software that is ready for delivery, and the user and maintenance manual for the software.

***

Reminder #176:

Before moving on to the next phase, conduct reviews on the project, i.e. plans, cost, schedule, risks, and perform re-estimation.

***

2.5 Delivery and Acceptance

Reminder #177:

Before doing any delivery of the software to the customer, test, test, and do more testing on the software.

***

Reminder #178:

The version of the software that is ready for release to the customer for acceptance should have already undergone and passed testing in the development environment, testing in the integration environment, and further reliability, duration and free-play test runs.

***

Reminder #179:

Invite software testers, external of the project team to perform testing on the software. These testing can give external insights on the readiness of the software for delivery.

***

Reminder #180:

Other than running through the test cases, allow for as much different free play test to be done on the software. This is sometimes a necessary evil to ensure that final software is sufficiently robust for operational usage.

***

Reminder #181:

Do not forget to record down the free play tests, so that if can errors or bugs occur, you have the steps that can be repeated for subsequent investigation and corrections. These free play test case can also be then added to the collection of test cases.

***

Reminder #182:

A test case must be linked to a requirement; if not there must be a description that explains the purpose of the test case.

***

Reminder #183:

The version of the software for release must be robust enough to allow free-play tests and take in all the possible abuse that an end-user can throw at the software without crashing the system.

***

Reminder #184:

Develop the software delivery and acceptance test plan as soon as the requirements are finalized, and have it reviewed internally, externally and with the customers.

***

Reminder #185:

With the software delivery and acceptance test plan, organize several meetings to explain to the customer the processes of how the software is released, and how the software will undergo the acceptance testing.

***

Reminder #186:

Treat all software delivery seriously. The software release must meet the customer's requirements, and must not have too many errors, such that it impacts the customer confidence on the software.

***

Reminder #187:

All software releases, internal or external, must be accompanied by a software release note that contains, at least, the version of the release, the name of the authorized staff issuing the release, what it contains, what issues it resolves from previous version (if any), what are the issues still outstanding (if any) and what are the enhancement (if any).

***

Reminder #188:

Develop, refine and test your software delivery processes as soon as possible, so that when its time of an actual software delivery, the process will be seamless.

***

Reminder #189:

The version of the source code and the binaries for all of the software release must be carefully controlled, and must be reproducible.

***

Reminder #190:

Assign a member of the project staff responsible for all version releases of the software. This staff will maintain all records of the software releases, and he/she must be part of the staff that will determine the fitness of the software for release.

***

Reminder #191:

The installation process of the software must be tested also before the software can be release.

***

Reminder #192:

Submit the acceptance test cases as soon as possible to the customer for their approval and acceptance. Organize meetings to explain to the customer the acceptance test cases and their links to the requirements.

***

Reminder #193:

Conducting user training and/or software maintenance training for the customer prior to the user acceptance tests, can give the customer the confidence and familiarity using the software. This helps to improve the standing of the software during the acceptance tests.

***

Reminder #194:

Acceptance test cases are important to software acceptance. They are usually the main mean to address and showcase to the customer that the software satisfy the software development project requirements.

***

Reminder #195:

Software developers must support the acceptance test activities. If there are any defects found during acceptance test, they must be available to correct the defects immediately.

***

Reminder #196:

Use an issue/defect tracking software tool to register and to keep track of all bugs and issues from the acceptance test. Monitor these until completion.

***

Reminder #197:

Categorize your bugs in terms of severity and priority; severity is defined as the level of impact a software bug has on the software, e.g. critical, major or minor and priority is defined the order of bugs need to be resolved, e.g. high, medium, or low.

***

Reminder #198:

Severity-priority combinations are:

(1) Critical-high,

(2) Critical-medium,

(3) Critical-low,

(4) Major-high,

(5) Major-medium,

(6) Major-low,

(7) Minor-high,

(8) Minor-medium, and

(9) Minor-low.

***

Reminder #199:

Software bugs that are classified severity critical and priority high would have implied that the software has failure in its basic functionalities. These must be rectified immediately.

***

Reminder #200:

At the other end of the spectrum, software bugs are classified severity minor and priority low. These can be scheduled for correction later, but these bugs still need to be scheduled for closure and closed.

***

Reminder #201:

Maintain key performance indicators to measure your progress, e.g. number of defects opened, number of defects closed, effort (days) per defects, the different categories of defects, etc. Also define the target that the project team needs to maintain or achieve and constantly share the updated key performance indicators with the project members and with the customers.

***

Reminder #202:

Ensure that all critical and major software bugs, issues and defects are corrected and completed before moving on to the next phase.

***

Reminder #203:

By the end of the delivery and acceptance phase, you should have completed acceptance tests with the customer, and the customers have accepted the software. The software may undergo operational trials after that.

***

Reminder #204:

Before moving on to the next phase, conduct reviews on the project, i.e. plans, cost, schedule, risks, and perform re-estimation.

***

Reminder #205:

If there are any outstanding bugs, issues or defects coming from the acceptance tests, endeavor and encourage your team to have those fixed, especially when you are left with those severity-minor and priority-low bugs or issues.

***

Reminder #206:

The completion of software acceptance tests is just only an episode of the software development project. This is usually followed on by post delivery activities, where the software is now being used in an operational environment.

***

Reminder #207:

After the acceptance tests, and with all the corrections completed and verified, the software now needs to be moved to the operational environment.

***

Reminder #208:

Your software delivery plan should include the activities describing how the software will be installed onto the operational environment and the testing activities there.

***

2.6 Post Delivery

Reminder #209:

Most software development projects come with 3/6/12 months of warranty period. This is the period where the software will be "tested" in an operational environment. The project team still needs to monitor the software usage, performance and reliability in the operational environment and until the project is completed (contractually).

***

Reminder #210:

You will need to be prepared (and your team also) for any software failure during the warranty phase. Make sure that faults, if any, are quickly registered, troubleshoot, the root cause(s) identified, and corrective action plans implemented.

***

Reminder #211:

You need a warranty/maintenance plan. This is the plan that says how the software will be supported during this phase. The plan usually includes SLA, KPIs, response time, turnaround time, software version update strategy, manpower allocation and support. Share this plan with everyone in the project, e.g. the customers and internally.

***

Reminder #212:

SLA stands for Service Level Agreements. SLA defines the services, e.g. scope, responsibilities, work description, etc, and the performance standard at which the defined services shall have. The SLA is usually part of the contract.

***

Reminder #213:

Understand and strictly abide to the turnaround time allowed for the different types (e.g. severity / priority) of issue/defect in the SLA. If there are any incidents that exceeded the time allow, investigate them immediate and address the root cause, as soon as possible.

***

Reminder #214:

All software faults reported must be registered on to an issue/defect tool, for tracking. These must be categorized in terms of their severity and priority. Track all these reports until closure.

***

Reminder #215:

Documents that need to be delivered:

(1) Software user documentation. This is the set of documents that describe how to use the software.

(2) Maintenance documents. This is the set of documents that describe how to perform the maintenance activities on the software.

***

Reminder #216:

You need a training plan. This is the document that describes objectives of the training, the different trainings required, how the trainings will be conducted, the duration of the trainings, the scope of training, the trainees' background requirements, etc. Share this plan with everyone in the project, e.g. the customers and internally.

***

Reminder #217:

Conduct Trainings:

(1) User training. This is the training that teaches the customers how to use the software.

(2) Maintenance training. This is the training that teaches the customers how to do maintenance on the software. The students of the maintenance training generally will become the level-1 support of the software.

***

Reminder #218:

Have frequent breaks in your user training so that the customer can digest the information presented. The breaks also give everyone a chance for better interactions.

***

Reminder #219:

In most instances, some of the personnel from the project team will continue the project into warranty and maintenance, where they will provide level-2 or level-3 warranty / maintenance support.

***

Reminder #220:

Level-1 support, or front end support, provides the initial support to the users of the software. The level-1 technician is generally located on-site, will generally record down the customer information, determine the customer issue, perform some analysis of the issue and if possible solve the problem if possible. If the issue cannot be addressed here, it is then escalated to level-2 support

***

Reminder #221:

Level-2 support provides a more in-depth support than level-1. The level-2 technician receives the reported issues from level-1, confirms the findings and further checks for the failing application, database, etc, and attempts to solve the issue here. If the issue cannot be addressed here, it is then escalated to level-3 support.

***

Reminder #222:

Level-3 is the highest level of troubleshooting. At this level, the software codes are analyzed with the data from the previous two levels. Sometimes, level-2 and level-3 are merged to form as a single level.

***

2.7 On Quality

Reminder #223:

QMS stands for Quality Management System. It is a set of policies, procedures, guidelines for the quality management of software products. Make use of standards such as ISO 91 and CMMI for the Quality Management System (QMS). Make sure that the QMS is suitable, can be implemented, and up to date.

***

Reminder #224:

You need to invest time, effort and money on quality. Quality does not happen from nowhere. Schedule these activities into your project schedule, where needed.

***

Reminder #225:

When implementing process improvement, avoid leap and bound implementation. Instead, improve your processes a step at a time, without fail and over a period of time. Practice makes perfect.

***

Reminder #226:

Educate all your project team members on ISO 91, CMMI and the QMS recommendations, and mention how these recommendations can help them and help the project.

***

Reminder #227:

When going for process certification, e.g. ISO 91 and CMMI, target ISO 91 first, CMMI level-2 second and follow by CMMI level-3. They are in the order of difficulty and maturity.

***

Reminder #228:

Instill quality awareness on the project team members. Quality induction must be conducted, whenever a new staff joins the software development project team. And constantly remind everyone on the quality

***

Reminder #229:

You need a quality assurance plan. This is the plan that describes that defines the definition of quality in the context of the software development project, how the quality of the software will be achieved, what are the measurements, and its supporting processes that will support achieving the software quality.

***

2.8 On Risks

Reminder #230:

Murphy's Law states that anything that can go wrong will go wrong.

***

Reminder #231:

Categorize your risks. If you do not have any categories yet, use the following and refine your risk categories as needed. :

(1) Contract,

(2) Requirements,

(3) Design,

(4) Test and Acceptance,

(5) Schedule

(6) Communication,

(7) Manpower,

(8) Quality,

(9) Processes,

(10) Change Control,

(11) Customer,

(12) Organization, and

(13) Others.

***

Reminder #232:

Develop a risk management plan that describes how risks will be anticipated, its impact estimated, and the project responses to the risks. Share this plan with your customer and your project team.

***

Reminder #233:

The software development project must have a risk list, and this risk list should be prioritized and know your top 10 risk without looking at your risk list. Constantly share the risk list with your customer and your project team. Keep the risk list and the risk management plan updated.

***

Reminder #234:

What you can do to a risk:

(1) Mitigate,

(2) Transfer,

(3) Accept, or

(4) Avoid

***

Reminder #235:

Conduct qualitative and quantitative risk analysis.

***

Reminder #236:

Your project needs to have a contingency budget for managing the risk and this budget ideally should not be lower that the risk amount exposed.

***

2.9 Others

Reminder #237:

Be a problem solver. Don't just talk about problems. Solve problems, think of alternatives, go and ask someone for help, brain-storm, and have some concrete action plans. Do something.

***

Reminder #238:

Understand and know the Pareto' principle or the 80/20 principle. This is a common rule of thumb. The Pareto principle states that 80% of the outcomes are the results of 20% causes, e.g. 80% of your project problems come from 20% of your project tasks.

***

Reminder #239:

When estimating, know that there are different types of estimating methods; expert judgment, by analogy, top-down estimating, bottom-up estimating, function point counting, COCOMO II, lines of codes, etc.

***

Reminder #240:

When estimating, be realistic in your estimation, and avoid doing (too much) padding. Padding is usually unjustified contingencies added to estimation.

***

Reminder #241:

An estimate will only be valid for a period of time. After that, the estimate will no longer be accurate. So, remember to revise your estimate when necessary, e.g. at the end of each phase of the software development project.

***

Reminder #242:

Perform at least 3 estimates, most likely, least likely, and average, and where possible perform the same estimation using different type of estimation methods.

***

Reminder #243:

In software development projects, you will probably need to estimate the cost of the project (and their sub costs), the duration and schedule of the project, the man-effort needed (in man-days), the size of the software.

***

Reminder #244:

But before escalating an issue, make sure that you have exhausted all your means to resolve the issue and record down these efforts down. And when escalating, make sure that you state what the problem is, what you have done, and what you need to solve the problem so that your bosses know how they can help you

***

Reminder #245:

If small milestones are constantly missed, especially at the start of the project, this is generally an indication that the project schedule is at a high risk of falling behind.

***

Reminder #246:

Instill in the project team to be serious with the project milestones and schedules at the start of the project. Any deviations from the plan must be justified with good reasons, and with corrective actions to prevent such deviations from happening again.

***

Reminder #247:

"Question: How does a project get to be a year late? ... One day at a time" by Frederick P Brooks.

***

Reminder #248:

Understand the rational that you need to be wary of estimates that claim to be 90% done, and why the last 10% may take as long the as 90%; ninety-ninety rule.

***

Reminder #249:

BOM stands for Bill of Material. Bill of material is a list of materials, with the quantities, needed to manufacture the end product. In the context of software, it is the list of software components that forms the final software.

***

Reminder #250:

When the project needs to purchase something, e.g. hardware, services, etc, get your procurement and logistics department involved and leverage on their expertise, to buy the things that the software development project needs. Make sure you adhere to your organization procurement processes when you need to purchase anything for your project.

***

3. Noteworthy References

1. Books

a. Project Management Metrics, KPIs, and Dashboards: A Guide to Measuring and Monitoring Project Performance 2nd Edition by Harold R. Kerzner (Author)

b. Excel Dashboards and Reports 2nd Edition by Michael Alexander (Author), John Walkenbach (Author)

c. A Guide to the Project Management Body of Knowledge: PMBOK(R) Guide 5th Edition by Project Management Institute (Author)

d. Software development project Survival Guide (Developer Best Practices) 1st Edition by Steve McConnell (Author)

e. Software Requirements 2 2nd Edition by Karl Wiegers (Author)

f. Project Management: A Systems Approach to Planning, Scheduling, and Controlling 11th Edition by Harold R. Kerzner (Author)

2. Websites:

a. Project Management Institute - http://www.pmi.org/

b. Microsoft Project - https://products.office.com/en-us/project/project-and-portfolio-management-software

c. Oracle Primavera - http://www.oracle.com/us/products/applications/primavera/p 6-professional-project-management/overview/index.html

d. Defect tracking software - https://www.atlassian.com/software/jira

e. Defect tracking software - https://www.bugzilla.org/

f. Standards - http://www.iso.org/iso/iso_90

g. Standards - http://cmmiinstitute.com/

***

About the Authors

C.W. OH and team advocate the need of adopting good processes for the successful completion of software development projects. OH worked as a project manager and had previously worked in project teams that delivered software systems to critical installations for private and government entities.

He can be reached at <u>c.w.oh.et.al@gmail.com</u>.

***

Index

***